EXECUTIVE SUMMARY

Throughout its fifty year history, Britain's nuclear industry has consistently failed to deliver on its promises. Now, less than five years after the financial collapse of British Energy, the UK's commercial nuclear generator, the public, parliament, and the financial markets are being asked once again to believe that a new generation of nuclear power stations can produce electricity safely and without government subsidy. And once again, there is good reason to believe that the industry's predictions are as spurious as in previous decades.

The general problems associated with nuclear power are well known (and largely unresolved). This report does not focus on these familiar issues but instead aims to provide the reader with an overview of the UK's nuclear industry and its history. We believe this to be essential in evaluating its proposals for a new generation of nuclear power stations in the UK. We hope that it will aid understanding for legislators, journalists and the general reader. It will also help in formulating strategy for campaigners and activists who wish to take action to stop the industry's plans.

Section 1 outlines the current political situation regarding the proposed new generation of power stations. We examine the positions of government and other major political parties and groups, the progress on new regulation being demanded by the industry, etc.

Section 2 presents a brief history of the UK's nuclear industry. It reveals its history of overestimating its capacities then consistently underperforming. We show that every nuclear power station constructed in Britain has suffered from time and budget overruns, producing less power than promised, with most of them ranked amongst the world's worst-performing.

Section 3 describes the current form of the UK's nuclear industry. We include some reflections on the strengths and weaknesses of the industry and its constituent companies.

Section 4 tells the story of troubled Olkiluoto 3 reactor, the only third generation nuclear power station in Europe. Currently under construction, in Finland, it is already far behind schedule and beset with difficulties.

Section 5 presents our conclusion that the industry is far weaker than it appears, with an array of unacknowledged weaknesses, and that the current overly optimistic proposal for ten new reactors is practically impossible.

Appendix 1 includes brief profiles of the main companies and organisations likely to be involved with any possible new power stations. We describe electricity producers, reactor designers, construction companies, regulatory bodies and other key organisations.

Appendix 2 gives sources of further information and campaign contacts

> ***"What exactly is nuclear power? It is a very expensive, sophisticated, and dangerous way to boil water."***
> *Helen Caldicott, author of Nuclear Power is not the Answer*

Section 1. Last Gasp or New Dawn for the Nuclear Age?

The British nuclear industry, like its counterparts in other countries, seemed, until recently, to be finished and headed for retirement, discredited by decades of spiralling costs, accidents and scandals, and the apparently insoluble problem of nuclear waste. No new reactors have been commissioned since the 1980s, and no one expected the ageing nuclear power stations to be replaced. However, an opportunistic and well-executed campaign has now forced nuclear power back onto the British political agenda.

The UK currently produces roughly 20% of its electricity from nuclear power stations and all but one of these is scheduled to close by 2023. Several of the most polluting coal plants must also shut: Britain is faced with a potentially crippling 'energy gap'. Now as oil and gas prices soar, domestic oil and gas reserves dwindle and pressure to tackle climate change grows, the nuclear industry has seized what may be its last opportunity to press for a new generation of reactors.

The Department for Trade and Industry may have been preparing for the return of nuclear power since at least 2002.[1] However, the government's 2003 Energy Review largely ignored the subject. Another review was soon ordered and published in July 2006. This latest review allows for new nuclear power if economically viable, leaving the decision up to the market. The review also recommends a 'streamlining' of the planning system and potentially other sweeteners to make nuclear power more economically viable, features lobbied for by the industry.

Why Nuclear? Why Now?

Its supporters rely on three main arguments in favour of nuclear power: that nuclear power produces low levels of carbon dioxide and is therefore necessary to help mitigate climate change; that uranium tends to come from politically stable countries and therefore represents security of energy supply; and that the looming energy gap that will be created by the closure of existing nuclear power stations, and many coal-fired power stations, can only realistically be filled by nuclear power.

These arguments depend on the assumptions that energy demand cannot realistically be reduced and that renewable power alone cannot produce enough reliable supply.

The pro-nuclear arguments are superficially plausible and backed up by a very well-funded public relations campaign. However, they do not stand up to close examination. The cross party House of Commons Environmental Audit Committee, for instance, has considered the matter in depth and their Sixth Report (published 28th March 2006) thoroughly refutes the arguments for new nuclear power. After taking evidence from expert witnesses from industry, NGOs, government and the science community, the committee reached some very clear conclusions: that the 'history of the nuclear industry gives little confidence about the time scales and costs of new build'; that 'nuclear can do nothing to fill the need for... new generating capacity... by 2016, as it simply could not be built in time'; that 'uranium mines can only supply just over half the current demand for uranium, and the situation is likely to become more acute'; whilst 'nuclear power can justifiably be regarded as a low-carbon source of electricity... the level of emissions associated with nuclear might increase significantly as lower grades of ore are used'; and that 'no country in the world has yet solved the problems of long-term disposal of high-level waste. The current work being conducted by CoRWM [the Committee on Radioactive Waste Management] will not be sufficient to address the issue.'[2]

The Next Steps Toward Nuclear

The DTI's Energy Review Team conducted consultations on the proposed legislative changes and aimed to produce white papers on energy and planning law in March 2007[3] . However, Greenpeace gained a judicial review of the consultation in the High Court, which found in Feb 2007 that the consultation process was fatally flawed and provided insufficient information. The government plans to issue a new consultation document, and publish its energy white paper in May 2007.[4]

Many of Britain's existing nuclear power stations were delayed by long planning inquiries hence the white paper aimed at streamlining planning law. The proposals in the Nuclear Policy Framework aims to reduce the freedom of planning inquiries to consider national regulatory and strategic questions; these would be established beforehand as would the licensing of reactor designs. Instead, the scope of inquiries would be restricted to local considerations.

According to the DTI:

> *The planning inquiry should focus on the relationship between the proposal and the local plans, and local environmental impacts. The inquiry should weigh up these local issues alongside the national strategic or regulatory material considerations, which will already have been established. The inquiry should also examine the local benefits of the development and how specific local impacts of the construction and operation of the plant can be minimised.*[5]

The proposed changes to the planning system are precisely what industry has been calling for. Bill Coley, CEO of British Energy, has described the current planning process as 'laborious'[6] and called for government to pre-license nuclear power station sites and fast-track planning permission for plants that meet generic, pre-agreed designs. In this way he argues that the construction times could be reduced to only four and a half years. His views are mirrored by EdF Energy and other leading voices elsewhere in the industry.

The nuclear lobby also wants 'a market mechanism... to complement the existing EU Emissions Trading Scheme and ensure a level playing field for low carbon technologies'. If carbon costs were to be fully accounted for and factored into electricity production, they believe that this would be sufficient to make nuclear power commercially viable.[7]

The Problem of Waste

One of the most intractable problems associated with nuclear energy is what to do with the waste products. In order to revive nuclear power it has been necessary for the government to devise a strategy to deal with nuclear waste. The Committee on Radioactive Waste Management (CoRWM), the government's committee appointed to make recommendations on waste strategy, produced its final recommendations only a few weeks after the energy review was released in July 2006.

The committee, appointed in 2004, was dogged by controversy throughout its operations. Few of its members are qualified engineers and scientists, and they have been accused of presenting data in a politically biased way. For instance, when they released figures on how much further waste might be produced by a new generation of reactors they emphasised the physical volume of waste that might be produced which they estimated at roughly 10% of the total volumes of existing waste. They glossed over the total radioactivity that would be produced, although their own figures showed an increase of roughly 165%; nearly tripling the radioactivity of Britain's waste stockpile[8]. They also made the assumption that fuel reprocessing, which massively increases the volumes of waste, would not be used, although reprocessing is standard practice in the UK. Furthermore, the estimates on waste from a new generation of reactors were based on the Westinghouse AP-1000 design, and were accepted uncritically from Westinghouse' then owners, BNFL (a company with a poor record for honesty and transparency). A CoRWM spokesman admitted that they 'didn't carry out much in-depth work on new build wastes.'[9] CoRWM's remit, set by government, also excluded consideration of low level waste, so the report ignores more than two million cubic metres of hazardous materials[10]; an amount which dwarfs the volumes of intermediate and high level waste.

Ultimately CoRWM's recommendation for waste storage deep underground are a postponement of action. Because local communities have always opposed any potential storage facility being imposed on them, CoRWM concluded that we must wait for a community to volunteer to host a facility. This shifts responsibility for waste storage away from government and onto the populace, and future generations. They estimate that it will take around forty years to complete the repository[11].

Other Political Parties

Nuclear power is a long term investment and to deliver a return would require a stable political climate in the long term. The positions of the other major political parties are therefore crucial. Industry does not want any surprises when a new government takes power. The Conservative party is apparently unenthusiastic: 'where the Government see nuclear power as the first choice, under our framework it would become a last resort; where the Liberal Democrats rule out nuclear power, we rule out subsidies and special favours for nuclear power.'[12] The Liberal Democrats are opposed to nuclear in principle considering it, 'a tried, tested and failed technology,'[13] and they actively campaign against it.

There are also several actively pro-nuclear trade unions. Prospect, for instance, which represents many nuclear workers, actively campaigns for nuclear power. In October 2006, Prospect met with industry leaders to 'debate the way forward for the delivery of new nuclear provision in the UK'; keynote speakers at the meeting included Bill Coley, CEO of British Energy, and Vincent de Rivaz, Chief Executive of EDF Energy[14]. Other unions in favour of nuclear power include the GMB, which considers that 'nuclear power has an important role to play as part of a balanced energy policy', and Amicus which 'welcomes the statement by the prime minister that we should replace our existing fleet of nuclear power stations'.[15]

Formerly known as the National Campaign for the Nuclear Industry, nUKlear 21 is a workers' campaign organisation with members from five unions, Amicus, GMB, Prospect, UCATT and TGWU. It actively campaigns for a new generation of nuclear power stations to be built. NUKlear 21 accepts money and other forms of support from BNFL.[16]

The nuclear debate seems also to have had a renaissance across Europe and there is high level support for it in Brussels. The European Commission published a strategic energy review on 10th January 2007[17]. Whilst the review said little about nuclear power, leaving the matter up to national governments, EU Energy Commissioner, Andris Piebalgs, has previously signalled a favourable attitude to nuclear, in language remarkably similar to Tony Blair's; 'We need to keep the choice of the nuclear option open for countries that want to generate electricity,' he said in a speech at a conference in Stavanger, Norway, and that 'nuclear energy presents the largest carbon-free energy source in the EU.'[18]

In Europe, there is currently a new reactor under construction in Olkiluoto, Finland (see section 5) with another planned for Flamanville, France. The nuclear resurgence is not confined to Europe. In the USA and Canada, and in the far east, in China and Korea, new reactors are under consideration. By some estimates the global market may require around 130 new nuclear reactors by 2020.[19]

Industrial Manoeuvres

Given government's insistence that there will be no subsidies for nuclear power it is unlikely that there will be any investment in it, unless industry's requests for some long term certainty (planning streamlining, carbon pricing) are granted. However, in a speech to the Utility Strategy Group, Vincent de Rivaz of EdF said that they will seek a licence to build a new £2.5bn nuclear power station in Britain early in 2007.[20]

German company E.On (owner of Powergen) has also announced its interest in building reactors in the UK. A member of the management board said that the company is 'looking to participate in the construction of new nuclear power plants in the UK'. Another German nuclear operator, RWE, is also interested in the British market.[21]

For new reactors to be built in the UK viable sites will be needed and it would be far more politically acceptable for new reactors to be built on existing nuclear sites than at new locations. Most of these are owned by the ill-fated British Energy (BE). There have been persistent rumours that French EdF Energy or German E.ON (owners of Powergen) may attempt to buy a stake. Utilities analyst, Lakis Athanasiou commented, 'The Germans and French are very keen on new build in this country. I have no doubts that behind closed doors, there's a lot of discussion going on between EdF, E.ON, BE and the government. BE has a big advantage in that it has the sites and the connections.'[22] In February 2007 BE announced that it is looking for partner companies to help build new reactors on its sites.[23]

In July 2006 the Department of Trade and Industry confirmed that the government was looking at selling its stake in BE. The sale might have raised £2 - 6 billion.[24] However, since then cracks in the graphite reactor cores and in reactor boiler tubes at several of BE's AGR power stations have caused the company to shut down a quarter of its electricity generating capacity, and the share price has plummeted by 30 percent. In November *The Independent on Sunday* reported that the government may delay the sale for a further two years.[25]

The government also plans to break up and sell BNFL Group's main subsidiary British Nuclear Group (see Section 4 BNFL). As this goes ahead it may have the effect of drawing many new private companies into the British nuclear industry, including many US based corporations which are keen to enter the lucrative decommissioning market.

The nuclear resurgence is not confined to Europe... by some estimates the global market may require around 130 new nuclear reactors by 2020

Section 2. Nuclear Construction in the UK, A Brief History

A total of nineteen nuclear power stations have been built in the UK on fourteen separate sites (see table 1)

Magnox - The first generation of British nuclear power stations were based on the Magnox reactor. The reactors were fueled by natural uranium contained in magnesium alloy canisters ('magnesium non-oxidising' hence Magnox). The reactors were moderated by a graphite core and cooled by pressurised carbon dioxide gas.

The world's first commercial nuclear power station was a Magnox station built by the UK Atomic Energy Authority (UKAEA) at Calder Hall, near Sellafield, Cumbria, and opened in 1956. A second was built at Chapelcross in Dumfries and Galloway, southern Scotland, and opened in 1959. Both these power stations were dual use: designed primarily to produce plutonium for nuclear weapons, generating electricity for the national grid was, at least initially, a byproduct.

A further nine Magnox power stations, all configured for energy generation over plutonium production, were ordered (by the Central Electricity Generating Board (CEGB) in England and the Southern Scottish Electricity Board (SSEB) in Scotland) between 1956 and 1963, and became operational between 1962 and 1971. Today the Magnox power stations, owned by the Nuclear Decommissioning Authority, are being retired; Berkeley in Gloucestershire was the first to close in 1989. Only two Magnox power stations remain operational, Oldbury and Wylfa, which are due to close by 2008 and 2010 respectively.[26]

AGR - The UK's second generation of nuclear power stations were based on the Advanced Gas Cooled Reactor (AGR) design. The AGR reactors were designed to be more efficient than the Magnox reactor and operated at higher temperatures, achieved by fuelling the reactor with enriched uranium oxide. Like the Magnox reactor the AGRs were moderated by a graphite core and cooled by pressurised CO_2. Seven AGR power stations were ordered by the CEGB and SSEB between 1965 and 1977 and became operational between 1976 and 1989. The AGRs, several of which have recently suffered extensive technical difficulties, are now owned by British Energy (BE) and are scheduled for closure between 2014 and 2023.[27]

PWR - Unlike the Magnox and AGR power stations both of which were based on uniquely British designs, a single power station, at Sizewell B in Suffolk, was based on Westinghouse' Pressurised Water Reactor (PWR). It was ordered in 1979 by the CEGB, and construction started in 1988; the plant became operational in 1995 and is forecast to close in 2035. The plant is now operated by BE.

> ***"The UK has never built a nuclear power station on schedule or within budget,"***
> *Walt Patterson, associate fellow in energy and the environment at the Royal Institute of International Affairs*[34]

Public Opposition

Throughout its early years in the 1950s and 1960s civil nuclear power in the UK faced little public opposition (especially when compared to that directed at nuclear weapons).

However, by the mid 1970s there was a growing public awareness of the dangers of nuclear power that made the building of new nuclear power stations increasingly difficult. Opposition to nuclear power was compounded by the nuclear accidents at Three Mile Island (1979) and Chernobyl (1986). From 1978 onwards the building of the last AGR at Torness, near Edinburgh, met with fierce local opposition including occupations of the construction site and numerous arrests[28]. The government's plan to build a fleet of new PWR power stations in the early 1980s met further opposition. For example, in 1981 the site of a proposed plant at Luxulyan Cornwall, was occupied by local people for six months. The CEGB were prevented from undertaking a preliminary survey of the site[29] and the plans were then dropped. The only proposed PWR plants to make any headway were Sizewell B and Hinkley Point C, both of which faced enormous public opposition and protracted public inquiries. The Sizewell B inquiry lasted for 340 days (1983-1985) at the time the longest public inquiry in British legal history[30]. The Hinkley C inquiry ran for thirteen months (1988-1989)[31]. Although planning permission was granted for both power stations only Sizewell B was built.

Construction

Construction of the UK's nuclear power stations was funded by the state-owned electricity generating utilities, the Central Electricity Generating Board (CEGB) and South of Scotland Electricity Board (SSEB). They were built by several consortia of electrical engineering, civil engineering and construction companies including Sir Robert McAlpine, Balfour Beatty and Taylor Woodrow [32]. Initially five consortia were created in 1955/6 to compete for contracts to design and build the first generation of Magnox power stations. By the time the first AGR power stations were ordered in 1965 the five consortia had been consolidated into three. The difficulties and delays encountered during the AGR building programme effectively bankrupted the remaining consortia.

The contracts awarded for both the Magnox and AGR construction programs gave responsibility for the design of entire power stations, including reactor design to the individual consortia. As a result, each consortium developed its own variation on the Magnox and AGR reactors. It was not until the later AGRs (for example, Heysham A and Hartlepool, and Hinkley Point B and Hunterston B[33]) that reactor designs were duplicated. Consequently there is far greater variation in UK reactor design than anywhere else in the world.

Time and Budget

Massive time and budget overruns, poor labour relations, and unexpected engineering and design problems were recurrent themes in nuclear power station construction in the UK. The first two Magnox plants, Bradwell and Berkeley, were completed over a year late[35]. The final Magnox plant, at Wylfa on Anglesey, ran three years over schedule[36].

This dismal pattern only deteriorated with the AGR power stations, none of which were completed on time or on budget. Dungeness B the first AGR plant to be ordered in 1965, and at the time expected to be operational in 1970/1971[37], did not produce commercial energy until 1989. According to the New Economics Foundation, Dungeness B exceeded its budget by 400%[38]. Problems persisted throughout the AGR building program: for example Heysham A was 13 years late (go ahead given 1970, expected operational in 1976[39], commercial operation started 1989), and cost almost double its original budget[40]. Even the final pair of AGR power stations, Torness and Heysham B, were both over a year late (approved in 1977, expected operational 1986/87, commercial operation started 1988). Torness was estimated to be 15% over budget[41].

Sizewell B was massively delayed by its planning inquiry. Initially ordered in 1979, construction did not start until 1988 and it produced commercial electricity in 1995. In 1982 the cost of the plant was estimated to be £2.01 billion; by the time of its completion costs had risen by almost 50% to £2.98 billion[42].

Power

As well as being late and over budget, the majority of the UK's nuclear power stations have also failed to perform to their designed output capacity (see table 2). For example, the AGR reactors, initially intended to operate at a load factor of 85%[43] have consistently failed to live up to expectations (see table 3). Most of the UK's reactors have performance figures in the lowest 25% of the world league table, with only two in the top 50% (see table 3). Even Sizewell B, built to an internationally standardised design, has a disappointing record compared to its counterparts in Germany and the USA[44].

> ***"The past history of the nuclear industry gives little confidence about the timescales and costs of new build. This does not mean that a new generation of nuclear power stations cannot be built to time and cost, but it does mean that investors have little basis for assessing the risks involved and may therefore require a higher rate of return"***
> *House of Commons Environmental Audit Committee, Sixth Report, March 2006*

TABLE 1. UK NUCLEAR POWER STATION PERFORMANCE[45]

Reactor	Type	Design Output (MW)	Authorised Output (MW)	Lifetime Load Factor	World Ranking (out of 414)
Dungeness A1	Magnox	275	225	59.2 %	366
Dungeness A2	Magnox	275	225	61.7 %	349
Oldbury 1	Magnox	300	217	57.8 %	313
Oldbury 2	Magnox	300	217	61.6 %	351
Sizewell A1	Magnox	290	210	57.4 %	375
Sizewell A2	Magnox	290	210	54.6 %	384
Wylfa 1	Magnox	590	490	59.5 %	363
Wylfa 2	Magnox	590	490	57.4 %	374
Dungeness B1	AGR	607	555	34.1 %	409
Dungeness B2	AGR	607	555	40.0 %	406
Hartlepool 1	AGR	625	605	56.8 %	379
Hartlepool 2	AGR	625	605	61.5 %	352
Heysham A1	AGR	611	575	58.1 %	371
Heysham A2	AGR	611	575	59.7 %	362
Heysham B1	AGR	615	625	74.0 %	189
Heysham B2	AGR	615	625	72.6 %	247
Hinkley Point B1	AGR	625	610	68.7 %	268
Hinkley Point B2	AGR	625	610	65.4 %	310
Hunterston B1	AGR	624	595	67.7 %	286
Hunterston B2	AGR	624	595	66.1 %	307
Torness 1	AGR	645	625	71.1 %	239
Torness 2	AGR	645	625	70.3 %	247
Sizewell B	PWR	1188	1188	83.5 %	49

TABLE 2. UK NUCLEAR POWER STATION KEY DATES

Name	County	Type	Current Status	Date ordered [46]	Est. finish date [47]	Construc -ction Start [48]	Commer-cial Power [49]	Date Closed [50]	Consor-tium [51]
Berkeley	Glos.	Magnox	Closed	1956	1961	1957	1962	2002	NEC
Bradwell	Essex	Magnox	Closed	1956	1960	1957	1962	2003	NPP
Calder Hall	Cumbria	Magnox	Closed	N/A	N/A	1953/ 1955	1956/ 1959	2004	UKAEA
Chapel-cross	Dumfries	Magnox	Closed	N/A	N/A	1955	1959	2006	UKAEA
Dungeness A	Kent	Magnox	Operational	1957	1963	1960	1965	2000	TNPG
Hinkley Point A	Somerset	Magnox	Closed	1957	1962	1957	1965	1990	BNDC
Hunter-ston A	Ayrshire	Magnox	Closed	1956	1961/ 1962	1957	1964	2008	AEG
Oldbury	Glos.	Magnox	Operational	1962	1966	1962	1967/ 1968	2006	TNPG
Sizewell A	Suffolk	Magnox	Operational	1961	1966	1961	1966	1991	BNDC
Trawsf-ynydd	Snow-donia	Magnox	Operational	1956	1961	1963	1971/ 1972	1989	APC
Wylfa	Angelsey	Magnox	Closed	1963	1968	1959	1965	N/A	BNDC
Dunge-ness B	Kent	AGR	Operational	1965	1970/ 1971	1965	1989	2018	APC
Hartlepool	Durham	AGR	Operational	1967	1974	1968	1989	2014	BNDC
Heysham A	Lancs.	AGR	Operational	1970	1976	1968/ 1970	1989	2014	BNDC
Heysham B	Lancs.	AGR	Operational	1977	1986/ 1987	1980	1989	N/A	NNC
Hinkley Point B	Somerset	AGR	Operational	1966	1972	1967	1976/ 1978	2011	TNPG
Hunter-ston B	Ayrshire	AGR	Operational	1967	1972	1967	1976	2011	TNPG
Torness	East Lothian	AGR	Operational	1977	1987	1980	1988/ 1989	2023	NNC
Sizewell B	Suffolk	PWR	Operational	1979	N/A	1988	1995	2035	NNC

Section 3. An Outline of the Nuclear Industry

For some companies, such as British Energy (BE) and BNFL, the delivery of nuclear power and related goods and services is their primary business. These companies are commonly thought of as 'the nuclear industry'. However, there are many other companies, essential to the operation of nuclear power stations, for which nuclear forms only a part of their portfolio of interests. In addition to nuclear fuels fabrication and nuclear electricity generation, the 'nuclear industry' overlaps several industrial sectors including: other electricity generation, engineering, and construction. A new generation of reactors would bring enormous volumes of new business for each part of the industry.

New plants would be commissioned by electricity generation companies. These companies own and operate nuclear plants and make their revenues (aside from subsidies) from selling the electricity they produce. Details of the electricity companies most likely to commission and pay for new nuclear plants in the UK can be found in Appendix 1. The most likely scenario for new plants is that they would be constructed on existing sites, owned by British Energy, in partnership with another generator, such as EdF Energy, E.On, or RWE. For the electricity generators a new generation of reactors represents continuity. For BE, which must shut most of its nuclear stations in the next 15 years, it would ensure survival as a large scale generator.

"the biggest problem facing the industry... a lack of trained staff from craft jobs right up to postgraduates because people have not been training nuclear engineers."

Nuclear power stations are vast construction projects taking years to complete, and requiring ongoing maintenance. For the construction and engineering companies the construction of up to ten new reactors would provide decades of big contracts. Ongoing maintenance over their projected 60 year lifespans and eventual decommissioning would total more than a century of new business. Most of the major construction companies are members of the Nuclear Industry Association. See Appendix 1 for further profiles of a few of the major UK construction companies with nuclear interests.

The companies which have developed third generation reactor designs are obviously key players in the nuclear industry. The front runners are Areva with their European Pressurised Water Reactor (EPR); Westinghouse, with the AP-1000 (Advanced Passive); and Atomic Energy of Canada Ltd, with the CANDU (CANada Deuterium Uranium) reactor (see Appendix 1 for further details of these companies). As well as the huge revenues that securing contracts in Britain would bring, the reactor designers and other specialist engineering companies have an eye on the potentially enormous worldwide market for new nuclear power stations . Projects in the UK would be an opportunity to develop and showcase new designs.

The industry also includes a group of state-owned companies, research and regulatory bodies, which provide essential services and, in the case of the regulators, give it legitimacy. Several of these, including the Nuclear Decommissioning Authority (NDA) and the UKAEA, are known to be lobbying for new reactors and BNFL is central to the industry's campaign. In Appendix 1 we include brief descriptions of the UKAEA, BNFL, the NDA, The Nuclear Installations Inspectorate (NII), Urenco, the Environment Agency, and the Health Protection Agency

Questions of capacity

There are doubts over the capacity of the British engineering and construction sector to deliver a new generation of nuclear plants. According to research by the Nuclear Industry Association (NIA), British industry presently lacks all the necessary capacities to construct and operate new reactors; the report finds that 'it would take around 5 years in the UK to get to the point where the industry could start construction'. But the industry has a deeper problem. Having failed to recruit adequately in recent decades, its workforce is ageing; 'many of the current experienced resources will be approaching retirement age over the next 5 to 10 years,'[52] according to the NIA.

In fact there is already a worrying shortage of trained staff to operate and decommission existing nuclear plants. The union Prospect calls it, 'the biggest problem facing the industry... There is a lack of trained staff from craft jobs right up to postgraduates because people have not been training nuclear engineers.' [53] Recruiting from abroad may not be easy either, with skills in high demand, as new nuclear power stations are commissioned around the world. The regulatory body, the NII, is also unable to recruit enough staff; it wants 180 inspectors but has only 165; 'we have had problems recruiting the right kind of people... [It is] an issue right across the industry'54, they said. In fact, the skills shortage is a problem across the whole engineering sector and engineers are in high demand worldwide.

The NIA believes that the situation can be remedied, 'by implementing a training programme to provide a new generation of nuclear technical specialists to fill the gap'. It does not appear to accept the possibility that the poor image of the industry has made it an unattractive career option.

Nuclear Spin

The industry also encompasses public relations and lobbying organisations. By necessity, it has long maintained a powerful political wing to maintain some semblance of public approval: a powerful coalition of trade associations, front groups, lobbyists and PR companies. Only a few years ago the industry seemed thoroughly discredited and headed for retirement. The speed with which it has regained credibility is a testament to the power of its public relations and lobbying.

For the nuclear lobby, a new generation of reactors would bring substantial volumes of work. There is no reason to imagine that nuclear power will ever cease to be controversial and its reliance on public relations is therefore likely to continue.

The NIA is the main nuclear lobby group in the UK, representing 130 nuclear industry companies. It plays a central role in the campaign for nuclear power. It works closely with BNFL, which provides more than half of its funding[55] and whose head of PR is chairman of the association. The NIA also works closely with the British Nuclear Energy Society. Another significant group is Supporters of Nuclear Energy (SONE) a campaign group run by Margaret Thatcher's former press secretary, Sir Bernard Ingham, who also campaigns against wind farms. Amongst other activities SONE is known to have lobbied the CBI for support in its campaign for nuclear new build[56].

For more details of the nuclear lobby and its work see, *www.nuclearspin.org* and *www.spinwatch.org.*

> ***"Nuclear technology, like all technology, is unavoidably subject to Murphy's Law: if something can go wrong it will. The engineers who design and build any complex piece of machinery always insist that it will be safe. They said that the Titanic, the Tay Bridge, and the Challenger space shuttle would be safe. After Windscale, Three Mile Island and Chernobyl, they still say other nuclear plants will be safe. The question is: who on earth believes them?"***
>
> *James Cutler & Rob Edwards,*
> *authors of Britain's Nuclear Nightmare*

Section 4. Case study: Olkiluoto, Prototype Nuclear Station in Finland

On the Baltic island of Olkiluoto, Finland, the first third generation reactor in the western world is under construction. Olkiluoto 3 is the prototype European Pressurised Water Reactor (EPR), and is being built by Areva and Siemens, next to two pre-existing nuclear power stations which were built in the 1970s and 80s[57].

The project progressed from inception to construction with remarkable speed. The electricity company Teollisuuden Voima Oy (TVO) applied for permission for the power plant in November 2000. Amid huge controversy, the Finnish parliament gave the go-ahead in May 2002, with Finnish public opinion evenly divided for and against the new reactor.

In December 2003 TVO signed a contract with a consortium of Framatome and Siemens (later Framatome ANP) for a 1600 MW EPR at a fixed price of €3.2 billion. TVO received permission to begin construction work in February 2005[58] . The Finnish Nuclear and Radiation Safety Authority, STUK, had taken only a year to consider the safety and suitability of the design. So certain was the consortium of receiving final permission, that excavation on the site and fabrication of the reactor vessel started before STUK had finished its studies.

Construction Delayed

Construction of the reactor began in early 2005, and although the design is much simpler than previous reactors, the project quickly ran into difficulties. By Spring 2006, the project was already delayed by around a year. By March 2007 it was a year and a half behind schedule[59]. There were problems in manufacturing reactor components; the quality of welding seams in the reactor pressure vessel were particularly problematic, failing quality control tests. There are also problems with the reactor base, where quality control failures have allowed the use of concrete which is too porous. Critics claim that the concrete porosity could increase corrosion of the concrete, weakening it [60]. Areva claims that this can be remedied by coating the concrete and that demolishing and rebuilding the base is unnecessary. The blame for these problems has been placed on the subcontractors, who were apparently unused to the demands of nuclear engineering projects.

The project is now a year behind schedule with the forecast date for commercial operation pushed back to 2010. Under the terms of the contract Areva will pay compensation to TVO for delays in the project. Areva and the French state aid agency will be the eventual payers. The expected date of commercial operations has now been pushed back to early 2011 and TVO estimates the cost of lost electricity production at around €600 million[61].

Financial Arrangements

The financial arrangements for the project are extremely unusual and, following a complaint from the European Renewable Energy Federation, the European Commission is investigating whether European trade rules have been breached[62]. The results of the EC investigation are not expected for at least a year.

One major anomaly is that it is a 'turnkey' contract, i.e. a fixed price contract for delivering a working power station on time. There are many inherent risks in big construction projects which often go way over budget and schedule due to factors which cannot be foreseen at inception.

Because it is a fixed price contract with penalty clauses for time overruns, TVO was able to secure a loan to help pay for the project at the extraordinarily low rate of 2.6%. Olkiluoto 3 is also effectively subsidised by a €570 million export credit arrangement between Areva and the French government agency Coface. Normally such insurance is reserved for deals with developing countries. The French taxpayer may therefore have to foot the bill.

TVO was able to get such a favourable deal due to intense competition for the contract amongst nuclear consortia. Being first with a functioning third generation reactor, enables Areva to showcase the EPR, whilst its competitors' designs remain just that, only designs. Areva was evidently prepared to pay for this advantage in a global reactor market which may demand over 100 reactors.

Safety Questions

Greenpeace commissioned two reports by nuclear consulting company John Large & Associates (L&A), in 2005 and 2006[63]. These reports looked at the safety of the EPR design and the approval process by STUK, and also at the implications of the problems with the concrete. L&A found that important features of the safety design were not finalised during the licensing, calling into question the licensing of the reactor, and the independence of STUK. It is also unclear whether the EPR can withstand a terrorist attack in which it could be struck by a passenger airliner.

The problem with concrete porosity could seriously impact on the durability of the reactor foundation slab and may require unapproved design changes. L&A also cast doubt on the capacity of STUK to effectively oversee the design and construction of the plant.

Conclusion

Olkiluoto 3 was intended to be a showcase for the claimed simplicity and economy of the third generation of nuclear power station. However, problems with the project - time and budget overruns, structural and safety problems - are all too reminiscent of previous generations of reactors. The streamlined design approval and planning processes should also be seen as a very pertinent warning of how new nuclear plants might be forced through in the UK.

SECTION 5. CONCLUSIONS

At first, the nuclear industry appears formidable. It includes many very large companies in several sectors and it has many powerful allies in government and across industry. The CBI, for instance, has been very supportive[64], as has the Energy Intensive Users Group, a heavy industry lobby group[65]. With its long history as a user of PR and lobbying services the industry has the sympathy of many within the PR sector, and its campaign for a new generation of reactors is well coordinated. Nuclear's comeback must be one of the most successful commercial propaganda campaigns of recent times, exploiting the sense of urgency about climate change and the looming energy gap. However, its carefully polished public image hides several serious weaknesses.

Although public acceptance has improved in the last few years with increasing awareness of the threat of climate change, public support for nuclear power is still not strong. Philip Dewhurst, chairman of the Nuclear Industry Association and head of PR at BNFL, said the campaign for nuclear power has 'largely been politically driven for the past three years...We need to win hearts and minds in local communities and reassure people about safety.[66]'

A recent MORI poll found that 39% of the British public support new build nuclear and 29% oppose it, with 32% undecided[39]. A survey of public attitudes by the University of East Anglia found that 34% of the public want nuclear power stations to be replaced as they reach the end of the lifespans; hardly massive support. In spite of its PR muscle, there is still widespread suspicion of nuclear power.

"Even as a Minister, let alone a Member of Parliament, I was never told the truth by the nuclear industry,"
Tony Benn

Broken promises

The British nuclear industry has several serious structural problems.

They would like to present nuclear power stations as reliable - quietly producing a constant electricity supply - and to contrast this with the intermittent nature of wind power. The truth as section 2 of this report reveals is quite different. Most of Britain's nuclear power stations have been plagued with stoppages and are far from reliable. The lifetime load factors for most of Britain's nuclear plants are very poor. Even Sizewell B, the umpteenth iteration of an internationally tried design of PWR, has poor figures compared to many others of the same design. The British nuclear industry has consistently failed to live up to its own hype. Not one British reactor was constructed on time and within budget. Most did not produce the promised electricity output either.

The industry's financial record is also very poor. The privatisation of nuclear power was shelved in 1989 and pushed through in 1996. British Energy collapsed only six years later in 2002 and in spite of high oil and gas prices is currently in deep financial difficulty again.

The industry has another even deeper structural problem. It has a lack of trained staff across the board and, having failed to maintain recruitment for many years, its workforce is ageing with large numbers of staff due to retire within ten years (see Section 3). They may struggle just to decommission existing reactors, let alone to build and operate ten replacement plants.

Taken together with the problems at Olkiluoto (see Section 5), the industry's record of broken promises and its ongoing problems give little confidence that it can deliver on its current pledges.

Vulnerability to public opposition

The campaign against nuclear power in the 1970s and 80s were extremely effective; even Margaret Thatcher's government, at the height of its power, managed to build only one nuclear power station: Sizewell B.

Now, as then, nuclear's power base is in central government and big business and it is over-reliant on misleading spin tactics to win public support. Already the government's public consultation process has fallen foul of the courts. Equivocal public support could easily melt away if nuclear power is once again perceived as an elite project, heavily spun and forced onto communities.

With liberalised electricity markets, in which consumers are free to switch electricity suppliers, generators with retail electricity operations (EdF Energy for instance) may also be vulnerable to consumer boycotts. Just as supermarkets were quickly forced to guarantee GM-free food, 'non-nuclear' electricity could become a selling point.

The construction companies that will be needed to build new reactors may also have particular vulnerabilities in that they have many thousands of sites across the country. Balfour Beatty, EdF Energy and RWE have a major public face in Metronet, for example, the joint venture responsible for maintaining and upgrading nine of the 12 London Underground lines. Construction generally has very slim profit margins compared to many industries, and if government stands by its word not to bail them out then disruption and delay due to protest could be very damaging financially.

Summary

If mistakes are not admitted and learned from, history tends to repeat itself. The British nuclear industry and its allies show no sign of having learnt anything from the hugely expensive errors of its past. We can be quite certain that if a new generation of British nuclear power stations is commissioned that budgets will mushroom and construction schedules will lengthen and future generations will inherit yet more of the lethal by-products of the technology.

Furthermore, current plans are unlikely by themselves to successfully plug Britain's looming energy gap or to deliver the required reductions in greenhouse gas emissions. Nuclear power is a red herring, distracting from the debate on these problems, and likely to divert the required energy and finances from effective solutions.

As before, the civil nuclear programme can be stopped. Strong though it is, the industry is brittle and has many vulnerabilities for its opponents to exploit. If current plans are defeated, history will show that the promised renaissance of nuclear power was in fact its last gasp.

> ***"From Windscale to Burghfield to Sizewell to Dounreay, there has been a catalogue of catastrophe, death and deceit. Accidents have been almost commonplace. Workers have been killed by radiation, the environment has been irrevocably poisoned with plutonium and innocent members of the public have suffered radiation related cancers."***
> *James Cutler & Rob Edwards,*
> *authors of Britain's Nuclear Nightmare*

Below we profile some of the key companies likely to be involved in new build nuclear stations. Most of the key companies in the

Appendix 1. Key Organisations and Companies

nuclear industry are members of the Nuclear Industry Association (NIA). A complete list of NIA member companies can be found at the NIA web site, www.niauk.org and at www.nuclearspin.org

Electricity Generators

British Energy - British Energy (BE) is the UK's primary nuclear generator. Whilst the old and inefficient magnox reactors were bequeathed to BNFL, BE operates the seven AGR stations and the one PWR at Sizewell B (see Section 2). As well as the 8 nuclear power stations (combined output - up to 9,600 MW) it operates one coal fired power station at Eggborough (2000MW) and also has a 50% stake (with AMEC) in the planned Isle of Lewis wind farm (702 MW). BE is the UK's largest electricity generator, producing roughly 20% of the UK's supply. In 2005/6 BE had revenues of £2.6 billion[68] .

British Energy Group comprises 27 companies including: British Energy Generation Ltd which owns and operates the nuclear power stations; British Energy Power and Energy Trading Ltd which manages all of the groups electricity sales; Eggborough which owns and operates the Eggborough coal plant; District Energy which runs four very small (10 MW) local gas-fired power stations; British Energy Direct which supplies electricity to over 2600 business customers, and has proven to be one of the most successful parts of the group[69]; British Energy Renewables which has stakes in a number of renewable energy projects; and a consultancy, British Energy NNC International Consulting (BENIC). BE employs 5,500 people[70].

British Energy Group Plc
Systems House
Alba Campus
Livingston
EH54 7EG
Phone: +44 (0)1506 408700
Fax: +44 (0)1506 408888
http://www.british-energy.com

The idea of privatising Britain's nuclear power stations was first suggested in the 1980s during the privatisation of the electricity sector. Although the last AGRs under construction were rushed to completion in 1989, the plan was abandoned as it was considered commercially nonviable. The nuclear plants were passed to two public companies Nuclear Electric and Scottish Nuclear[71] which were later merged as British Energy, and finally privatised in 1996, after a lengthy PR and lobbying campaign coordinated by the PR company Hill & Knowlton[72].

In 2002 when electricity prices reached a record low the company collapsed. The government bailed it out with loans and took a 65% share in the company, which it would now like to sell, possibly to EdF or E.On.

The sale could have raised between £2 and £6 billion[73]. However, despite rising energy prices and record profits in 2006 BE has run into financial difficulty again. Total energy production has fallen and BE has been plagued by problems with its ageing and unreliable AGRs. This will be the third consecutive year in which production targets will be missed[74]. After problems with cracks in the graphite reactor cores and in reactor boiler tubes at several power stations, the company has shut down a quarter of its electricity generating capacity and the share price has plummeted by 30 percent[75]. Government plans to sell its shares have been shelved.

British Energy is a member of the NIA and the company secretary, Robert Armour, is a member of SONE.

EDF Energy - EDF (Electricite de France) is a French based power utility. It has a near monopoly on power generation and supply in France. The company was created in 1946 with the nationalisation of the French power industry[76]. EDF was partially privatised in November and is now only 87.3% owed by the French state[77]. As well as its French operations EDF operates in most European countries as well as Africa (Ivory Coast and Egypt), the Americas (Argentina, Brazil, Mexico and USA) and Asia (China and Vietnam). The company had a turnover of £33.2 billion in 2004[78].

Electricité de France
22-30, avenue de Wagram
75382 Paris Cedex 8, France
Phone: +33-1-40-42-22-22
Fax: +33-1-40-42-79-40

http://www.edf.fr

EDF is one of the largest electricity and gas supply companies in the UK. It is represented by EDF Energy which bought the power supply companies, London Energy, SWEB and Seeboard Energy[79]. Most of the electricity generated by EDF in the UK comes from coal and gas fired power stations[80]. Much smaller amounts come from combined heat and power plants and renewable sources[81].

EDF Energy
40 Grosvenor Place
Victoria
London
SW1X 7EN
Tel: 020 7242 9050
info@edfenergy.com

EDF operates all of France's nuclear power stations a total of 58 reactors on 19 sites supplying 75% of French electricity[82]. It also has a 50% stake in Belgian nuclear power station Tihange-1[83].

EDF is actively pushing for new nuclear power stations worldwide. It is preparing to build a new nuclear power plant at its existing site of Flamanville in northern France[84]. If completed, the plant will be the first new nuclear power station to be built in France for over ten years. EDF is also a member of NuStart, a consortium set up to push for the building of new nuclear power stations in the USA[85]. EDF also wants to build nuclear power stations in China[86].

EDF is one of the loudest industry voices calling for the building of new nuclear power stations in the UK. In November 2005, EDF's Chief Executive, Vincent de Rivaz, told a parliamentary select committee that EDF could build new nuclear power stations within ten years if planning and licensing laws are streamlined[87]. EDF says it can help to make nuclear acceptable to public opinion[88]. The Chancellor Gordon Brown's brother, Andrew, is head of media relations at EDF Energy[89].

E.ON - E.ON AG is a German based power and gas company. The company was formed in June 2000 by the merger of VEBA and VIAG, two of Germany's largest industrial groups with interests in power generation, mining, oil, telecommunications and chemicals[90]. The company now focuses on power and gas and has operations throughout Europe and in the United States[91]. E.ON had a turnover of € 56.4 billion in 2005[92].

E.ON
E.ON-Platz 1
40479 Düsseldorf
Germany
T +49 2 11 - 45 79 - 0
F +49 2 11 - 45 79 - 5 01
E-mail: info@eon.com
Internet: www.eon.com

E.ON have four businesses in the UK: Central Networks (electricity distribution formerly Midlands Electricity and East Midlands Electricity), E.ON Energy (power and gas for businesses), E.ON UK (power generation and supply) and Powergen (domestic and business power supply)[93]. Over 90% of E.ON's UK power generation capacity comes from gas, coal and oil power stations, a further 6% comes from combined heat and power plants and just over 2% from renewable sources[94].

E.ON UK
Westwood Way
Westwood Business Park
Coventry
CV4 8LG
England
T +44 24 - 76 42 - 40 00
F +44 24 - 76 42 - 54 32
E-mail: enquiries@eon-uk.com
Internet: www.eon-uk.com

E.ON Kernkraft is the largest private operator of nuclear power stations in Germany. They currently own and operate eight of Germany's seventeen operational nuclear power stations (Brunsbüttel, Unterweser, Krümmell, Grohude, Brohdorf, Isar 1 and 2, and Grafenrheinfeld) and have a minority stake in a further two (Lingen (12.5%) and Gundremmingen (25%))[95]. E.ON also have a minority stake in British/Dutch/German-owned nuclear fuel company Urenco[96].

E.ON has expressed an interest in building new nuclear power stations in the UK. Like the other nuclear generators, E.ON wants regulatory concessions on planning, and carbon pricing[97]. E.ON has expressed interest in building a further nuclear power station in Finland[98].

RWE AG - RWE or Rheinisch-Westfälische Elektrizitätswerk AG is a German utilities company. RWE was founded as an electrical power company in 1898[99] and is now one of the biggest diversified utilities companies in the world. It has interests in the electricity, gas and water sectors and operates in Czech Republic, Germany, Hungary, Netherlands, Poland, Slovakia, UK and USA[100]. RWE had a turnover of £27.8 billion in 2005[101].

RWE AG
Opernplatz 1
45128 Essen, Germany
Phone: +49-201-12-00
Fax: +49-201-12-15199

RWE's major UK company is npower[102]. npower is one of the UK's largest suppliers of electricity and gas. It owns electricity and gas brand names MEB, Calortex, Independent Energy, York Gas, Yorkshire Electricity and Northern Electric[103]. The vast majority of power generated by npower comes from coal, gas and oil fired power stations[104]. It also operates a number of combined heat and power plants[105]. npower owns Juice, a renewable energy company, launched in conjunction with Greenpeace UK, with its power coming from npower's North Hoyle Offshore Wind Farm[106].

RWE npower
Trigonos
Windmill Hill Business Park
Whitehill Way
Swindon
Wiltshire
SN5 6PB
Tel: +44(0)1793 877777

Until October 2006, RWE also owned Thames Water, the largest water company in the UK[107], and the world's third largest water supplier with operations in Africa, Asia, Australia, Europe, and South America[108]. Thames Water became hugely controversial in 2006 for generating massive profits for RWE, whilst its ageing pipe network was leaking and a drought order was declared[109]. Thames Water has been sold for £8 billion to a consortium led by the Australian bank, Macquarie[110].

Of the 17 nuclear power plants currently operational in Germany, RWE owns the two nuclear power plants at Biblis[111], near Frankfurt, and has a majority stake in the power two plants at Gundremmingen, near Stuttgart, (75%)[112] and the single plant at Lingen, on the German/Dutch border (87.5%)[113]. RWE have recently sold their nuclear engineering subsidiary company RWE NUKEM[114] to Advent International[115]. Its unclear whether the RWE minority stake in nuclear fuel company Urenco[116] was included in the sale to Advent International or is retained by RWE.

Construction Companies

AMEC - In recent years AMEC has grown from a major UK construction company to a multinational employing 45,000 people in 40 countries, with revenues of £2.6 billion in 2005 with pre-tax profits of £124 million[117]. It works in the oil and gas, transport, power and utilities and nuclear sectors. Its major divisions working in the nuclear industry are AMEC Nuclear and AMEC NNC, formerly the National Nuclear Corporation.

AMEC has a major strategic interest in the nuclear industry, and has even previously expressed an interest in buying British Nuclear Group, the main part of BNFL[118].

Whilst most of the large British construction companies would be attracted by large contracts in the sector, nuclear seems to play a more central role in AMEC's business model. AMEC has major interests in nuclear power in Canada, the USA and South Africa as well as in Eastern Europe and countries of the former Soviet Union. It is engaged in major engineering and project management services to the global nuclear industry as well as decommissioning and waste treatment[119].

AMEC Nuclear
Booths Park
Chelford Road
Knutsford
Cheshire
WA16 8QZ
Phone: +44 (0) 1565 633800
Fax:
www.amec.com, www.amecnnc.co.uk

AMEC is a member of the NIA and non-exec director Neville Chamberlain is also a member of SONE. The pro-nuclear former Energy Minister Brian Wilson is also a non-exec director of AMEC Nuclear[120].

AMEC is no stranger to controversy, having involved itself in the controversial Baku-Ceyhan pipeline which runs from the Caspian Sea to the Mediterranean and is linked to extensive human rights abuses and environmental problems[121]. AMEC also built the Birmingham Northern Relief Road and has major investments in many PFI projects.

Balfour Beatty - Balfour Beatty is the largest construction company in the UK, with revenues of £4.9 billion in 2005 and profits of £115m. BB employs 27,000 people worldwide. Within it almost 200 subsidiary companies work in four major areas of business: building, civil and specialist engineering, rail engineering and investments[122].

Though BB is not a member of the NIA, as Britain's largest "concrete pourer", it can be expected to tender for contracts on new nuclear power stations. The subsidiary BBCEL has undertaken some major contracts as part of the decommissioning of the Hunterston A magnox reactor, including the construction of a store for intermediate level waste there and also at Dounreay and Sellafield[123].

Balfour has a 20% share in Metronet, a joint venture with Atkins, Bombardier, Thames Water (RWE), and EdF Energy[124] Metronet is responsible for the maintenance, renewal and upgrading infrastructure on nine of the London Underground lines[125].

Major shareholders of Balfour include Aviva plc with 5.99%, Legal and General with 3.02%, Morley Fund Management with 5.0%, Prudential with 3.49%, Royal and Sun Alliance with 4.05%, and Standard Life with 3.83%[126].

Balfour Beatty plc
130 Wilton Road
London
SW1V 1LQ
Phone: +44 (0)20 7216 6800 or 6801
Fax: +44 (0)20 7216 6902
http://www.balfourbeatty.com

In 2002, Balfour Beatty and a number of other major European companies were forced out of the failed Ilisu dam project in Turkey[127]. Extensive pressure was brought to bear on the company, whose share of the project was worth £200m. The Ilisu Dam Campaign harnessed a broad coalition of NGOs and campaigners focussing on the environmental and human rights implications of the proposed dam. Amongst other tactics the campaign successfully lobbied several institutional investors in the company to support an NGO-sponsored motion at the company's AGM which would have forced the company to withdraw from the project. Though the motion failed, the unprecedented interest shown by investors is thought to have forced Balfour out of the Ilisu project.

Carillion - Carillion is a large UK construction company, formed when Tarmac Construction de-merged from the Tarmac group in 1999. It has been involved in nuclear construction work since the 1950s. It built the Chapelcross Magnox power station, and has also worked at the Dounreay experimental reactor site and on many projects at Sellafield[128]. Mowlem, with which Carillion recently merged, also has a long history in the nuclear industry[129]. Carillion has completed 36 PFI projects and has many more planned or under construction.

Turnover in 2005/6 was £1.01 billion. Interim results for 2006/7 were much higher due to acquisition of Mowlem. The company employs 40,000 people. Carillion is a member of the NIA

Carillion plc
550 Mauldeth Road West
Chorlton-cum-Hardy
Manchester
M21 7EE
Phone: +44 (0) 161 860 2056
Fax: +44 (0) 870 128 4816

In the 1990s, as Tarmac Construction, the company built the M3 extension through Twyford Down, the M65 at Stanworth Valley and (in partnership with AMEC) the second runway at Manchester Airport. All were the subject of major campaigns. The company responded to public criticism with one of the first major greenwash campaigns, publishing the booklet 'Tarmac in the Environment'[130] and attempting to rebrand itself as an environmentally aware company. Despite the rebranding, Tarmac still made a (failed) bid to construct the Newbury bypass, although they claimed that they would not have built the road without major design changes to mitigate environmental damage.

Sir Robert McAlpine - Sir Robert McAlpine (SRM) is a large British civil engineering company. Founded more than 130 years ago and based in Hemel Hempstead, its turnover for 2005/6 was £782 million with a profit of £13.8 million[131]. It has offices throughout the UK and many offices in Europe and beyond. SRM should not be confused with Alfred McAlpine a separate construction company, mostly involved in house building.

SRM is ultimately owned by members of the McAlpine family.

Sir Robert McAlpine Ltd
Eaton Court
Maylands Avenue
Hemel Hempstead
Hertfordshire, HP2 7TR
Tel. 01442-233444

Fax. 01442-230024
SRM has a long history of involvement in Britain's nuclear power programme, joining a consortium, The Nuclear Power Group[132] in 1966. In all it has worked on the construction of six nuclear power stations, including Hinkley Point B, Hunterston B, and most recently, Torness[133] and Sizewell B.

SRM has been involved in many other nuclear installations, including the MOX fuel plant at Sellafield (an addition to the THORP reprocessing plant) and several other major projects there. The MOX plant was due for completion in July 1995, however it was not completed until 1997 and did not begin operations until after the millennium. The costs were forecast at around £300 million[134] but ended up at £473 million[135].

SRM is a member of the Nuclear Industry Association and Sir William McAlpine is a member of Supporters of Nuclear Energy (SONE).

Other Construction Companies

Other large construction companies who are members of the Nuclear Industry Association, and can therefore be expected to bid for contracts on new build nuclear, include: Laing O'Rourke, the largest privately owned construction company in the UK - the previously independent Laing, worked on the construction of Sizewell B in the 1980s; Taylor Woodrow - whilst predominately a housebuilder, Taylor Woodrow has been involved in nuclear power construction since the magnox reactor programme; and Kier Group, another major construction company.

Nuclear Engineering and Design Companies

Areva - Areva is a French energy company, over 96% owned by the French government[136]. Areva had 2005 sales of US$11.99 billion[137]. It was formed in September 2001 with the merger by Commissariat a L'Énergie Atomique (CEA - the French Atomic Energy Commission) of its nuclear fuel and reprocessing company Cogema, its nuclear reactor company Framatome and its electrical transmission company FCI[138]. Today Areva has three operating divisions reflecting the original companies: Areva NP, develops and builds nuclear reactors and is 34% owned by German conglomerate Siemens; Areva NC is a nuclear fuel company, involved with the whole nuclear fuel cycle, from mining to waste disposal; and Areva T&D is a power transmission and distribution company.

Areva is a globally important nuclear company. Its subsidiary Areva NP is a major player in nuclear reactor design and construction. Originally called Framatome, Areva NP began as a joint venture between French engineering company Schneider Group and American reactor builder Westinghouse[139]. It was responsible for the design and construction of all of France's second generation of nuclear power stations (PWR reactors). Framatome has also exported its PWR reactor technology to Belgium, South Africa, South Korea and China[140]. Areva claim to have built 98 of the world's 443 nuclear reactors [141] Framatome ANP was created in 2000 by the merger of Framatome and the nuclear reactor business of Siemens[142].

Areva NC is a major global player in nuclear fuel production and waste reprocessing. The company extracts 15% of the world uranium supply and operates uranium mines in Niger, Canada and Kazakhstan[143]. The company operates the EURODIF uranium enrichment facility in south-eastern France[144], which supplies nuclear fuel worldwide, and also operates the nuclear reprocessing facility at La Hague, in northern France, which reprocesses nuclear waste from all over the world, and produces Mixed Oxide (MOX) nuclear fuel. Greenpeace New Zealand have highlighted the large amounts of weapons usable plutonium produced by La Hague (nearly 80,000kg by the end of 2000)[145].

Areva
27 - 29 Rue Le Peletier
75 433 - Paris cedex 09 - France
Tel.: 33 (0)1 44 83 71 00
Fax: 33 (0)1 44 83 25 00

Areva's current UK activities are confined to Areva T&D's work in electricity transmission and delivery. They have recently won a major contract with EDF Energy for upgrading electricity substations in London[146].

Areva T&D UK Ltd
PO Box 27
Lichfield Road
Stafford
ST17 4LN
Staffordshire
Tel: 01785 257111
Fax: 01785 252540
www.areva-td.com

EPR [147]- One of the leading reactor design contenders for UK new build nuclear power stations is the European Pressurised Water Reactor (EPR). The EPR, effectively a next generation of PWR, is being developed by Areva NP. The first EPR is currently under construction at Olkiluoto in Finland (see Section 4) and construction of a second may start soon at Flamanville in France. Areva is lobbying hard in both China[148] and USA[149] to build further EPR based power stations.

Westinghouse - US based Westinghouse Electric Company provides fuel, services, technology, plant design, and equipment to the nuclear electric power industry[150]. Its AP-1000 reactor design is one of the leading designs suggested for the next generation of UK reactors Whilst it was until recently owned by BNFL, Westinghouse has only one site in the UK[151]., the Springfields nuclear fuel fabrication facility near Preston, Lancashire.

Springfields Fuels Limited
Westinghouse
Springfields
Salwick
Preston
PR4 0XJ
Tel.: +44 (0) 1772 762000

Founded in 1886 by George Westinghouse, the company entered the nuclear power market in 1947, designing and building the US' first commercial nuclear power station at Shippingport, Pennsylvania, which opened in 1957[152]. Westinghouse went on to develop the Pressurised Water Reactor (PWR), firstly for military ships and submarines, and later as a civil power source. Westinghouse' first PWR opened in 1960 at Yankee Rowe in Massachusetts[153].

Westinghouse is one of the principle nuclear power station designers in the world. The company claims 'Westinghouse PWRs represent 76 percent of the light water reactors around the world; 67 percent of the PWRs are based on Westinghouse PWR technology'[154].

Between 1972 and 1981 Westinghouse owned 45% of French reactor builder Framatome which had been licensing Westinghouse reactor technology since 1958[155]. It spun off many of its core businesses from the late 1980s onwards[156] and invested heavily in media companies. In 1995 Westinghouse acquired the media giant CBS and in 1997 the entire company adopted the CBS name. Viacom then bought CBS in 1999, and sold its nuclear power operations to British Nuclear Fuels Limited (BNFL), rebranded as Westinghouse Electric Company. In 2000 the nuclear power businesses of Swiss engineering company ABB were also purchased by BNFL and integrated into Westinghouse Electric Company[157].

BNFL has now sold the company to Toshiba, completing the sale on October 16th 2006, of a 77% share in the company[158]. Toshiba is already a leader in boiling water reactors and believes that the acquisition of Westinghouse and its PWR technology will give them a major advantage in the resurgent nuclear energy market which they expect will require more than 100 new reactors worldwide by 2020.

AP-1000 - Westinghouse is offering a third generation design of civil power reactor called AP-1000, one of the most likely designs to be used in the UK. The AP-1000 is a redesign of the company's AP-600 and is named for its Advanced Passive (AP) safety features and its greater than 1000 MWe output. The AP-600 was licensed for use in the USA in the 1990s but none were built as they would not have proven competitive. The passive safety features which would shut down the plant in the event of an accident use gravity and are judged to be safer than powered systems[159].

Whilst no AP-1000s have yet been built, the design has obtained design certification from the U.S. Nuclear Regulatory Commission and may be used in 10 proposed U.S. atomic plants, including one at Bellefonte power station in Hollywood, Alabama, the site of two partially constructed PWR reactors, mothballed in 1988. Four AP-1000 reactors are also expected to be built in China[160]. It is the favoured design of the NuStart Energy nuclear power consortium, one of the three major nuclear power consortia to have formed in the US in recent years[161].

Atomic Energy of Canada Ltd (AECL)

AECL is the major nuclear operator in Canada, and is also interested in winning contracts in the UK. It conducted a study of new reactors with British Energy in 2003. In Januray 2007, it sent a delegation to meet with the UK's nuclear regulators and the DTI in order to promote its latest CANDU reactor design, the ACR-1000[162]. The CANDU reactors have been under development since the late 1950s.

Other nuclear engineering companies

Other companies which specialise in nuclear engineering include: Jacobs (formerly Jacobs Babtie), a very large and highly diversified engineering, design and consultancy company working in many different engineering sectors, it conducts design and consultancy for most of the major nuclear companies in the UK; Doosan Babcock (previously Mitsui Babcock), an energy services engineering company describing itself as 'the largest supplier of operational support for the nuclear power generation sector in the UK'; and Atkins Nuclear, the nuclear engineering arm of consultancy Atkins.

Nuclear Fuel and Services

BNFL Group - No survey of the British nuclear industry could be complete without some consideration of BNFL. One of the most controversial companies in the UK, BNFL is central to the British nuclear industry. It is the largest funder of the Nuclear Industry Association and has pushed strongly for a new generation of nuclear plants in the UK.

BNFL Group (formerly British Nuclear Fuels Ltd) was created in 1971 by separating out the fuel fabrication division of the UKAEA. It is now a holding company whose activities 'span the entire nuclear energy cycle, from reactor design and fuel manufacture to power station decommissioning and clean-up.'[163] The group comprises British Nuclear Group and Nexia Solutions. The reactor design business Westinghouse (see Westinghouse section) was recently sold to Japanese conglomerate Toshiba[164]. The company had a turnover of £2.496 billion in 2005/6[165] and employed around 23,000 people worldwide. With the sale of Westinghouse and the transfer of magnox reactors to the Nuclear Decommissioning Authority (NDA), turnover in 2006/7 will be much lower, and the company now employs around 13,400 people[166].

British Nuclear Group (BNG) manufactures and reprocesses nuclear fuel, operates the UK's old magnox reactors (although ownership of the magnox reactors has been transferred to the NDA) and also conducts nuclear decommissioning and clean up operations[167]. BNG's chief site is the controversial Sellafield complex on the Cumbrian coast.

Nexia Solutions is a nuclear technology company which provides support services to nuclear plant operation, decommissioning, nuclear policy development, and laboratory management[168].

BNFL's Sellafield complex (formerly Windscale) has a long history of radioactive accidents[169]. Most recently, British Nuclear Group was fined £500,000 by the Health and Safety Executive following the leak of highly radioactive liquid waste in the THORP plant discovered in April 2005[170].

BNFL
Daresbury Park
Warrington
WA4 4GB
Phone: +44 (0) 1925 832000
Fax: +44 (0) 1925 822711

Plans to privatise the BNFL Group have been proposed and abandoned many times. The decommissioning and clean up contracts associated with the break up could be hugely lucrative.

The American conglomerate Fluor has expressed an interest in buying BNG whole. Other potential buyers, include a consortium of Serco (which manages AWE Aldermaston) and Amec[171]. The deal is mired in controversy over whether to break the group up and by disputes with the Nuclear Decommissioning Authority. The union Amicus, which represents many nuclear industry workers, has called for the government to sack the management of both the NDA and BNFL whom it accuses of gross incompetence and conflicts of interest. Doug Rooney of Amicus said, 'my concern is that there will be a stitch-up between the parties. They are looking to satisfy the self-interest of managers who have demonstrated that they are inept.'[172] Amicus insists that any sale must be open to competition.

The recent plans to sell BNG failed in Autumn 2006 and the company is now to be divided up and profitable elements sold off piecemeal[173]. On 6th February 2006, BNFL announced the sale of British Nuclear Group Project Services Ltd, a specialist nuclear decommissioning and clean up division of BNG[174].

Urenco - Urenco (uranium enrichment company) produces enriched uranium for nuclear fuel[175]. The company is one third British, one third Dutch, and one third German. The British third is state-owned through BNFL. The Dutch third is largely owned directly by the Dutch government (98.9%) with a tiny stake owned by Shell, Philips, DSM, and VMF-STORK. The German third is owned by RWE and E.On[176].

Urenco Ltd
18 Oxford Road
Marlow
Buckinghamshire
SL7 2NL
Phone: +44 (0) 1628 486941

Public sector organisations

UKAEA - Founded in 1954 to oversee the development of both nuclear weaponry and nuclear power, the United Kingdom Atomic

Energy Authority initially oversaw every part of the UK's nuclear programme[177]. Over its history many of its functions have been hived off into other national bodies or privatised. Fuel production was demerged in 1971 to form the BNFL and Urenco companies[178]. In 1973 the nuclear weaponry part became the Atomic Weapons Establishment. In 1982 the Amersham centre for radiochemistry research was privatised as Amersham International (now Amersham plc)[179]. In 1996 many more commercially viable elements were spun off as AEA Technology.

Now the UKAEA, based at Harwell in Oxfordshire, conducts research into nuclear fusion (at the Joint European Torus, Culham, Oxfordshire) and decommissions nuclear sites around the country, working as a contractor to the Nuclear Decommissioning Authority.

Nuclear Installations Inspectorate

The Nuclear Installations Inspectorate (NII) is the major part of the Nuclear Safety Directorate (NSD) a division of the Health and Safety Executive (HSE) and is responsible for enforcing the laws on nuclear installations. In practice, the names NII and NSD are used interchangeably. The NII monitors and enforces all health and safety regulations on nuclear sites[180].

In 2001, British Energy and BNFL asked the NII to start a pre-licensing review for the AP1000, however the NII declined due to a lack of adequate staffing[181]. The HSE/NII is currently preparing guidance for the licensing of new reactor designs. They envisage that the licensing process will take at least three and a half years[182].

Independent nuclear consultant, John Large, has accused the NII of complacency in inspecting British Energy's fleet of ageing reactors as they apparently did not notice cracks in the boiler tubes and graphite cores until the problem had become serious enough to force the closure of several plants[183].

Environment Agency

Whilst the NII is responsible for workers' safety on site at nuclear installations, the Environment Agency is responsible for the enforcement of laws against pollution of the wider environment. It is responsible for the licensing and monitoring of all kinds of radioactive pollution and waste. It licenses organisations using radioactive substances and sets the emissions limits for nuclear installations, as well as conducting some inspections and monitoring of sites[184], and also monitors radioactivity in the environment. The EA's two Nuclear Regulation Groups employ no more than 30 individuals to oversee environmental matters at the UK's 33 major nuclear installations[185].

Health Protection Agency, Radiation Protection Division
(formerly National Radiological Protection Board)

The National Radiological Protection Board, was merged with the Health Protection Agency, on 1st April 2005, forming the Radiation Protection Division, based at Chilton, Oxfordshire[186]. This is itself a subset of the HPA's Centre for Radiation, Chemical and Environmental Hazards. The Radiation Protection Division undertakes research into radiation risks and protection from them, provides training and lab services and is an important advisory body for UK policy on radioactivity. It also reviews the work of other committees, for instance that of the Committee Examining Radiation Risks of Internal Emitters (CERRIE).

Nuclear Decommissioning Authority

The Nuclear Decommissioning Authority (NDA) is a new public body set up in April 2005 to oversee and issue contracts for the clean up of many of the UK's nuclear sites. These include the old magnox reactors as well as many UKAEA research sites, such as Winfrith in Dorset and Dounreay, Caithness. The transfer of ownership of the magnox sites from British Nuclear Group to NDA was highly controversial. It could set a precedent for nuclear companies off-loading their clean up liabilities without paying the full cost. However, the transfer was cleared by the European Commission in April 2006, which had investigated whether it constituted illegal state aid[187].

References

1 - Rob Edwards, 'Secret plan to revive UK nuclear power industry', New Scientist, 3-8-2002
2 - House of Commons Environmental Audit Committee, Sixth Report, 28-3-2006, http://www.publications.parliament.uk/pa/cm200506/cmselect/cmenvaud/584/58402.htm
3 - 'First concrete could be poured for a new nuclear unit', NucNet, 12-2-2007; "Energy generators tire of being kept in dark", Financial Times, 12-10-2006
4 - Christopher Adams, 'UK nuclear plans delayed until May', Financial Times, 22-2-2007
5 - DTI web site: www.dti.gov.uk/energy/review/implementation/nuclear-framework/page31831.html
6 - Guy Dixon, 'Nuclear boss urges end to 'laborious' planning process', Scotland on Sunday, 27-8-06
7 - British Energy submission to Energy Review, April 2006
8 - CoRWM's 'Radioactive Waste and Materials Inventory - July 2005', www.corwm.org.uk/pdf/1279%20-%20Task%20088%20CoRWM%20Inv%20July%202005%20Final.pdf
9 - telephone interview with Steve Mansfield, CoRWM, 4th September 2006
10 - CoRWM's 'Radioactive Waste and Materials Inventory - July 2005', www.corwm.org.uk/pdf/1279%20-%20Task%20088%20CoRWM%20Inv%20July%202005%20Final.pdf
11 - CoRWM 'Final Report', Chapter 16, paragraph 55, p132 http://www.corwm.org/pdf/FullReport.pdf
12 - 'Interim findings of the Conservative Party's Energy Review', 6-7-2006, http://www.qualityoflifechallenge.com/index.php/archives/interim-findings-of-the-conservative-partys-energy-review/
13 - 'New nuclear build must not be foisted on communities', Liberal Democrat Party website, http://www.libdems.org.uk/news/story.html?id=11019&navPage=news.html
14 - "New nuclear build: How do we make it happen?", Prospect press release, 2-10-2006, www.prospect.org.uk/news/newsstory.php?news=377
15 - Stakeholder responses to nuclear power proposals, EPolitix.com, 17-5-2006, www.epolitix.com/EN/ForumBriefs/200605/62c8556f-caf9-4d56-80ff-6cdff04d2e87.htm
16 - Nuklear 21 profile, Nuclearspin.org, www.nuclearspin.org/index.php/NUKlear21#The_Concept_of_.22Greenhouse_Free.22
17 - Dieter Helm, 'Europe's Energy Future: in the dark', 16-1-2007, www.opendemocracy.net/globalization-institutions_government/europe_energy_4251.jsp
18 - Stephen Voss, 'EU Energy Commissioner backs nuclear power, to aid environment',Bloomberg, 23-8-06, www.bloomberg.com/apps/news?pid=20601085&sid=ax5tfBNbBsw4&refer=europe
19 - Toshiba Press Release 17-10-2006 "Toshiba Completes Westinghouse Acquisition"
20 - Richard Orange, "EdF plans to lead Britain into the new nuclear age", The Business, 25-11-2006
21 - NucNet, 24-11-06, www.worldnuclear.org
22 - Guy Dixon, 'British Energy poised to profit from next generation game', Scotland on Sunday, 20-8-2006
23 - Terry Macalister, 'British Energy calls for partners to build new wave of nuclear plants', Guardian, 14-2-2007
24 - Guy Dixon, 'British Energy poised to profit from next generation game', Scotsman, 20-8-2006
25 - 'British Energy sell-off frozen for two years - report', Hemscott, 2006, http://www.hemscott.com/news/latest-news/item.do?newsId=37754910018052
26 - British Nuclear Group website http://www.britishnucleargroup.com/section.php?pageID=186
27 - British Energy website http://www.british-energy.com/pagetemplate.php?pid=82
28 - Walt Patterson, "Going Critical: An unofficial history of British nuclear power", 1985, page 48, available online at http://www.waltpatterson.org/goingcritical.htm; also Scottish CND webpage on Torness Power station available online at http://www.banthebomb.org/archives/scotland/toracc.shtml
29 - "R v Chief Constable of the Devon and Cornwall Constabulary, ex parte Central Electricity Generating Board, Court of Appeal, Civil Division, [1982] QB 458, [1981] 3 All ER 826, [1981] 3 WLR 967, 146 JP 91, hearing dates: 22, 23, 24, 25 Sept, 20 Oct 1981, 20 Oct 1981" available online at http://www.hrcr.org/safrica/arrested_rights/R_Chief%20Constable.htm
30 - British Energy press release, "Sizewell B power station marks 10 years", available online at http://www.britishenergydirect.com/article.php?article=23, and "Nuclear power generation development and the UK industry" available online at http://www.dti.gov.uk/energy/sources/nuclear-power/technology/generation/page17922.html
31 - Chris Alder, "Hinkley Fears Rise", Somerset County Gazette, 18-05-06, available online at http://www.somersetcountygazette.co.uk/display.var.764009.0.0.php
32 - The Monopolies And Mergers Commission, "Central Electricity Generating Board: A Report on the operation by the Board of its system for the generation and supply of electricity in bulk", 1981, appendix 25 p359-61
33 - Adrian Ham and Robert Hall, 'A way forward for nuclear power', 2006, p 43, available online at http://www.dti.gov.uk/files/file28276.pdf
34 - Cole Moreton, 'Going nuclear: News Analysis', The Independent, 21/05/06
35 - Walt Patterson, 1985, "Going Critical: An unofficial history of British nuclear power", p.7, available online at http://www.waltpatterson.org/goingcritical.htm
36 - Walt Patterson, 1985, "Going Critical: An unofficial history of British nuclear power", pp.7-8, available online at http://www.waltpatterson.org/goingcritical.htm
37 - The Monopolies And Mergers Commission, "Central Electricity Generating Board: A Report on the operation by the Board of its system for the generation and supply of electricity in bulk", 1981, p.35, table 3.2, available online at http://www.competition-commission.org.uk/rep_pub/reports/1981/136cen_elec_gene_board.htm
38 - Andrew Simms, Petra Kjell and David Woodward, "Mirage and Oasis: Energy choices in an age of global warming", 2005, p34, available online at http://www.neweconomics.org/gen/z_sys_PublicationDetail.aspx?PID=209
39 - The Monopolies And Mergers Commission, "Central Electricity Generating Board: A Report on the operation by the Board of its system for the generation and supply of electricity in bulk", 1981, appendix 25 p355, available online at http://www.competition-commission.org.uk/rep_pub/reports/1981/136cen_elec_gene_board.htm
40 - Paul Brown, 'Mistake shuts down new nuclear reactor', The Guardian, 15-8-1984
41 - James Buxton, 'Torness Reactor Produces Power For The First Time', Financial Times, 26-3-1988, p.3
42 - 'Sizewell B - first or last of a kind?', FT Energy Newsletters/Power Europe, 10-2-95, p.2
43 - "The best reactors in the world?", FT Energy Newsletters/Power Europe, 8-3-1996, p6
44 - Steve Thomas, 'The economics of nuclear power: analysis of recent studies', July 2005, p.8, available online at http://www.psiru.org/reports/2005-09-E-Nuclear.pdf
45 - ibid. Table 5, p.7 and Table 6, p.8
46 - Adrian Ham and Robert Hall, 'A way forward for nuclear power', 2006, appendix A, tables A1, (p34) and A2 (p43) and A3 (p46) available online at http://www.dti.gov.uk/files/file28276.pdf
47 - various sources including:
'Bradwell reactor takes shape', The Times, 20-3-1958
'Two nuclear plants at Hinkley Point', The Times, 12-4-1960
'Nuclear Power Station Delay - Change in design at Hinkley Point', The Times, 29-6-1960
'Nuclear Power: Siting and Safety", Stanley Openshaw, 1986, table 4.1, p.128
'Going Critical: An unofficial history of British nuclear power", Walt Patterson, 1985
48 - International Atomic Energy Authority Power Reactor Information System database available online at http://www.iaea.org/programmes/a2/index.html
49 - International Atomic Energy Authority Power Reactor Information System database available online at http://www.iaea.org/programmes/a2/index.html
50 - British Energy and Nuclear Power Group websites
51 - The Monopolies And Mergers Commission, 'Central Electricity Generating Board: A Report on the operation by the Board of its system for the generation and supply of electricity in bulk', 1981, p.358, APPENDIX 26
(Referred to in paragraph 12.93) available online at http://www.competition-commission.org.uk/rep_pub/reports/1981/fulltext/136appendices.pdf
52 - Nuclear Industry Association, 'The UK capability to deliver a new nuclear build programme',pp 22, 31, www.niauk.org/pdf/MAIN_REPORT_12_march.pdf
53 - Terry Macalister, 'British nuclear renaissance faces threat of skills meltdown', Guardian, 31-6-2006
54 - ibid.
55 - Chris Grimshaw, 'Paying to be Propagandised', Corporate Watch Newsletter 32, Oct 2006; based on interviews, press searches and documentation received from BNFL under Freedom of Information rules
56 - ibid.; see also www.sone.org.uk
57 - http://www.ol3.areva-np.com/project/index.htm
58 - http://www.olkiluoto.info/en/
59 - Satu Hassi, 'Nuclear flagship project in Finland faces major problems', Inter Press Service News Agency, March 2007
60 - Ariane Saines, 'Results of STUK's Olkiluoto-3 probe expected at end of April', Nucleonics Week, vol 47, number 10, 9-3-2006, p4-5, http://www.platts.com/Content/Nuclear/Newsletters%20&%20Reports/Nucleonics%20Week/See%20A%20Sample/index.pdf
61 - "Further delay in construction of Olkiluoto-3 nuclear reactor", Helsingin Sanomat, 5-12-2006, http://www.hs.fi/english/article/Further+delay+in+construction+of+Olkiluoto-%203+nuclear+reactor/1135223464390
62 - Tobias Buck, 'French loan for nuclear reactor under scrutiny', Financial Times, 25-10-2006, http://www.ft.com/cms/s/e92124c0-63c5-11db-bc82-0000779e2340.html
63 - John Large & Associates web site: www.largeassociates.com/3149%20Olkiluoto/R3149-A1%20Final%20Issue.pdf; www.largeassociates.com/R3123-a2%20final%20Issue.pdf
64 - ibid.; also for example, CBI press release, 29-11-2005, www.cbi.org.uk/ndbs/press.nsf/0/617504617c5f9adc802570c8004e6ffe?OpenDocument
65 - EIUG Response to DTI Energy Review Consultation, 13-4-2006, www.eiug.org.uk/publics/l1304w1.PDF
66 - Ravi Chandiramani, 'Nuclear sector hunts for public champion', PR Week, 22-6-2006
67 - MORI, 'Attitudes to Nuclear Energy', 7-12-2006, http://www.ipsos-mori.com/polls/2006/niauk2.shtml
68 - www.british-energy.co.uk/pagetemplate.php?pid=77, also from FAME database of companies data
69 - British Energy Group - Final Results - Part 1, 20-6-2006
70 - NIA profile, www.niauk.org
71 - Steve Thomas, 'The Economics of Nuclear Power', July 2005, p.22
72 - Hill & Knowlton, web site, no longer available
73 - Guy Dixon, 'British Energy poised to profit from next generation game', Scotsman, 20-8-2006
74 - ibid.
75 - 'British Energy sell-off frozen for two years', Hemscott, 2006, www.hemscott.com/news/latest-news/item.do?newsId=37754910018052
76 - Wikipedia, en.wikipedia.org/wiki/Électricité_de_France
77 - EdF web site, actionnaires.edf.com/75155i/Homecom/Shareholders/StockMarket/Shareholdingpolicy.html
78 - Hoovers compaby profile, www.hoovers.com/electricité-de-france/--ID__92402--/freeuk-co-factsheet.xhtml
79 - Wikipedia, http://en.wikipedia.org/wiki/EDF_Energy
80 - EdF web site, http://www.edfenergy.com/html/showPage.do?name=edfenergy.energy.til
81 - EdF web site, http://www.edfenergy.com/html/showPage.do?name=edfenergy.energy.til
82 - 'Nuclear Notes from France', web site of the French Embassy in the USA, Spring 2003, http://www.ambafrance-us.org/intheus/nuclear/n2f2/spring2003.asp
83 - EdF web site, www.edf.com/96157i/Homefr/EDFInternationalpresence/inEurope/Belgium/EDFBelgium.html
84 - EdF web site, http://www.edf.fr/html/epr/uk/index.html
85 - NuStart Energy consortium web site, http://www.nustartenergy.com/Consortium.aspx
86 - 'EdF lauds China stance on nuclear power plant', People's Daily Online, 17-12-2003 http://english.people.com.cn/200312/17/eng20031217_130586.shtml
87 - House of Commons Environmental Audit Committee, Environmental Audit - Minutes of Evidence, 2-11-2005, www.publications.parliament.uk/pa/cm200506/cmselect/cmenvaud/584/5110201.htm
88 - Michael Harrison,"France's EDF wants to build nuclear power stations in UK", Independent, 2-11-2005

89 - Terry Macalister, "The powerful business of promoting a nuclear future: Chancellor's brother among successful lobbyists for new generation of power plants", The Guardian, 11-7-2006
90 - E.On web site, www.eon.com/en/unternehmen/2061.jsp;jsessionid=127CE7F7D63A3405E47AE8B7483AED79.2
91 - E.On web site, http://www.eon.com/en/unternehmen/2056.jsp
92 - E.On web site, http://www.eon.com/en/unternehmen/2040.jsp
93 - E.On web site, http://www.eon-uk.com/34.aspx
94 - calculated from figures given on E.On web site, http://www.eon-uk.com/933.aspx
95 - 'Nuclear Power in Germany', Nuclear Issues Briefing Paper 46, Feb 2007, www.uic.com.au/nip46.htm; www.kellogg.northwestern.edu/faculty/diermeier/ftp/atom/exhibit2.htm
96 - 'Urenco Comnpany links', WISE web site, 31-5-2006, http://www.wise-uranium.org/ecure.html
97 - E.On web site, paras 21-32 http://www.eon-uk.com/libraries/uk/downloads/whoweare/E.ON_UK_energy_review_response.pdf#search=%22eon%20%22energy%20review%22%22 (10/09/06)
98 - http://business.guardian.co.uk/story/0,,1890709,00.html (09/10/06)
99 - http://www.rwe.com/generator.aspx/rwe-group/chronicle/language=en/id=6132/1898-1920-page.html (10/09/06)
100 - use 'please select a country' option http://www.rwe.com (10/09/06)
101 - http://www.hoovers.com/rwe/--ID__42385--/freeuk-co-factsheet.xhtml (10/09/06)
102 - http://www.rwe.com/generator.aspx/country/grossbritannien/language=en/id=160820/page.html (10/09/06)
103 - http://www.rwe.com/generator.aspx/rwe-npower/group-structure/language=en/id=231742/rwe-npower-npower.html (10/09/06)
104 - http://www.rwe.com/generator.aspx/rwe-npower/group-structure/language=en/id=231398/rwe-npower-generation-and-renewables.html (10/09/06)
105 - http://www.rwe.com/generator.aspx/rwe-npower/group-structure/language=en/id=231398/rwe-npower-generation-and-renewables.html (10/09/06)
106 - http://www.npower.com/at_home/juice-clean_and_green/about_juice/how_juice_started.html (10/09/06)
107 - www.thameswater.co.uk/en_gb/Downloads/PDFs/History_of_Thames_Water_new.pdf (10/09/06)
108 - http://www.waterjustice.org/uploads/attachments/pdf80.pdf (10/09/06)
109 - Jonathan Prynn, 'Thames Water boss is paid Pounds 8million... while leaking London', Evening Standard, 22-6-2006
110 - Mark Milner, 'Thames Water sold for £8bn to Australian bank Macquarie', Guardian, 17-10-2006
111 - Daniel Diermeier, 'Nuclear power plants in Germany', www.kellogg.northwestern.edu/faculty/diermeier/ftp/atom/exhibit2.htm
112 - Wikipedia (German language), http://de.wikipedia.org/wiki/Kernkraftwerk_Gundremmingen
113 - RWE Power, 'A brief portrait: Lingen power plants', no date available, http://www.rwe.com/generator.aspx/rwe-power-icw/standorte/konvent-kraftwerke/erdgas/emsland/informationsmaterial/property=Data/id=109366/emsland-2-engl-pdf.pdf#search=%22lingen%2087.5%22
114 - NIA profile, http://www.niauk.org/pdf/RWE.pdf
115 - Nukem web site, http://www.nukem.co.uk/
116 - 'Urenco company links', WISE web site, http://www.wise-uranium.org/ecure.html
117 - Amec, 2005 Annual Report, http://www.amec.com/investors/investorsreports.asp?pageid=28&mryear=2005
118 - Charlotte Moore, "Government to sell British Nuclear Group", The Guardian, 30-3-2006
119 - Amec web site, http://www.amec.com/sectors/sectors_2ndlevel.asp?PageID=264
120 - Amec web site, www.amec.com/news/mediareleasedetails.asp?Pageid=876&MediaID=1079&mryear=2005
121 - Baku Ceyhan Campaign web site, www.bakuceyhan.org.uk
122 - from FAME database of company data, and balfourbeatty.co.uk
123 - Balfour Beatty web site, www.balfourbeatty.co.uk
124 - Douglas Friedli, 'Balfour Beatty profits stay on course to rise despite Tube train headache', Scotsman 13-8-06
125 - Metronet web site, www.metronetrail.com/
126 - from FAME database of companies data
127 - Ilisu Dam Campaign web site, www.ilisu.org.uk
128 - NIA profile, www.niauk.org
129 - 'Carillion's approach to the nuclear sector', Carillion web site, www.carillionplc.com/assets/documents/powerpoint/nuclear_sector.pdf
130 - 'Company Statements on Sustainable Development', 19-11-1999, Biothinking web site, http://www.biothinking.com/applysd/statements.htm
131 - from FAME database of companies data
132 - Monopolies and Mergers Commission, 'Central Electricity Generating Board: A Report on the operation by the Board of its system for the generation and supply of electricity in bulk', 1981
133 - Sir Robert McAlpine web site, www.sir-robert-mcalpine.com/projects/index.html
134 - John Vidal, 'The Last Blast', The Guardian, 16-12-1993
135 - Angela Jameson, 'Sellafield faces an uncertain future beyond the clean-up', The Times, 31-5-2003
136 - Areva Resources Canada web site, http://www.cri.ca/contact/recent_questions.php
137 - Hoovers proifle of Areva, http://www.hoovers.com/areva/--ID__104852--/free-co-factsheet.xhtml
138 - 'Cogema becomes Areva NC', Areva web site, http://www.areva-nc.com/servlet/ContentServer?pagename=cogema_en%2FPage%2Fpage_html_libre_full_template&c=Page&cid=1039482706542
139 - 'Nuclear power in France - why does it work?', Nuclear Power Corporation of India Ltd web site, http://www.npcil.nic.in/nupower_vol13_2/npfr_.htm
140 - 'Nuclear Power in France', Uranium Information Centre web site, http://www.uic.com.au/nip28.htm
141 - 'Areva publishes submission to UK Energy Review', Areva web site, 17-5-2006,
www.areva.com/servlet/ContentServer?pagename=arevagroup_en%2FPressRelease%2FPressReleaseFullTemplate&cid=1146746863523
142 - Siemens profile, Reaching Critical will web site, www.reachingcriticalwill.org/corporate/dd/DDpdf/sie.pdf#search=%22framatome%20%20westinghouse%20history%22
143 - Cogema web site, http://www.cogema.com/servlet/ContentServer?pagename=cogema_en%2FPage%2Fpage_html_libre_full_template&c=Page&cid=1143726745175
144 - Cogemaweb site, http://www.cogema.com/servlet/ContentServer?pagename=cogema_en/site_prod/site_prod_full_template&c=site_prod&cid=1039473237746
145 - 'No Nukes', Greenpeace New Zealand web site, www.greenpeace.org.nz/campaigns/nuclear/areva.asp
146 - 'Areva wins 12 million euro contract', Electric Energy Online, 20-6-2005, www.electricenergyonline.com/IndustryNews.asp?m=1&id=37844
147 - Wikipedia, http://en.wikipedia.org/wiki/European_Pressurized_Reactor
148 - 'Foreign energy giants bid for China's nuclear contracts', People's Daily Online, 12-9-2004, english.people.com.cn/200409/12/eng20040912_156691.html
149 - Press release, Unistar Nuclear web site, 15-9-2005, www.unistarnuclear.com/09-05release.html
150 - Westinghouse web site, http://www.westinghousenuclear.com/Our_Company/index.shtm
151 - telephone interview with Alan Beauchamp, Employee and Media Communications, Westinghouse, 20-10-2006
152 - Westinghouse history available online at http://www.westinghousenuclear.com/Our_Company/history/Timeline/1900_1939.shtm and
http://www.westinghousenuclear.com/Our_Company/history/Timeline/1940_1979.shtm and http://www.westinghousenuclear.com/Our_Company/history/Timeline/1980_2003.shtm
153 - 'Outline history of nuclear energy', Uranium Information Centre web site, www.uic.com.au/nip50.htm
154 - Westinghouse web site, http://www.ap600.westinghousenuclear.com/F2.asp
155 - Carole Collins, 'Framatome: French nuclear monopoly finds fertile ground abroad', Multinational Monitor, vol.4 no.7, July 1983, multinationalmonitor.org/hyper/issues/1983/07/collins.html;
156 - For example see http://www.westinghouse.com.au/home (viewed 19/10/06) a domestic appliance brand owned by Electrolux
157 - Westinghouse web site, www.westinghousenuclear.com/Our_Company/history/Timeline/1980_2003.shtm
158 - 'Toshiba Completes Westinghouse Acquisition', PR Newswire, 17-10-2006, www.prnewswire.com/cgi-bin/micro_stories.pl?ACCT=no&TICK=WE&STORY=/www/story/10-17-2006/0004453746&EDATE
159 - 'Building The Next Generation Of Reactors, The Chief Engineer, www.chiefengineer.org/content/content_display.cfm/seqnumber_content/2671.htm
160 - 'Shaw gets 20 percent of Westinghouse', M&C News, 4-10-2006, news.monstersandcritics.com/business/article_1208137.php/Business_Roundup
161 - 'Grand Gulf, Bellefonte and River Bend chosen for COL applications', Nuclear Engineering International, 24-9-2005, www.neimagazine.com/story.asp?storyCode=2031440
162 - Rebecca Bream, 'Nuclear groups plan for contracts ahead of energy white paper', Financial Times, 30-1-2007
163 - BNFL web site, www.bnfl.com/content.php?pageID=7
164 - Reuters 30-8-2006
165 - BNFL Annual Report 2006, p.48
166 - BNFL web site, www.bnfl.com/content.php?pageID=7
167 - British Nuclear Group web site, www.britishnucleargroup.com/content.php?pageID=23
168 - Nexia Solutions web site, www.nexiasolutions.com/section.php?pageID=23
169 - see for instance James Cutler & Rob Edwards, "Britain's Nuclear Nightmare", 1988, especially chapter 2
170 - HSE press release, E103:06, 16-10-2006, www.hse.gov.uk/press/2006/e06103.htm
171 - Tom McGhie, "Serco bids to keep nuclear group British", Mail on Sunday, 3-9-2006
172 - Angela Jameson, 'Amicus accuses BNFL of Magnox stitch up', The Times, 12-10-2006
173 - Katherine Griffiths, 'Anger as nuclear sell-off is shelved', Telegraph, 23-8-2006
174 - British Nuclear Group web site, http://www.britishnucleargroup.com/content.php?pageID=31&nID=2105
175 - Urenco web site, www.urenco.com
176 - 'Urenco company links', WISE web site, www.wise-uranium.org/ecure.html
177 - UKAEA web site, http://www.ukaea.org.uk/about/our_history.htm
178 - Wikipedia, http://en.wikipedia.org/wiki/Bnfl
179 - Amersham plc web site, http://www.amersham.com/about/heritage.html
180 - phone interview with HSE spokesman, 25-10-2006
181 - Parliamentary Office of Science and Technology, Dec 2003
182 - HSE web site, http://www.hse.gov.uk/nuclear/reactors/proposals.htm
183 - File on Four, "Heading for a Nuclear Blackout?", BBC Radio 4, 5-12-2006
184 - Environment Agency web site, www.environment-agency.gov.uk/business/444304/945835/?version=1&lang=_e
185 - Environment Agency web site, www.environment-agency.gov.uk/business/444304/945835/1064129/1085181/?lang=_e
186 - Health Protection Agency web site, www.hpa.org.uk/radiation/
187 - David Gow, "EC paves way for UK nuclear privatisation", The Guardian, 5-4-2006